MEASURING VOLUME

by Meg Gaertner

Cody Koala
An Imprint of Pop!
popbooksonline.com

abdobooks.com
Published by Pop!, a division of ABDO, PO Box 398166, Minneapolis, Minnesota 55439. Copyright © 2020 by POP, LLC. International copyrights reserved in all countries. No part of this book may be reproduced in any form without written permission from the publisher. Pop!™ is a trademark and logo of POP, LLC.

Printed in the United States of America, North Mankato, Minnesota

102019
012020

THIS BOOK CONTAINS RECYCLED MATERIALS

Cover Photo: Shutterstock Images
Interior Photos: Shutterstock Images, 1, 19; iStockphoto, 5 (top), 5 (bottom left), 5 (bottom right), 7, 9, 11 (top), 11 (bottom left), 11 (bottom right), 12–13, 15 (top), 15 (bottom left), 15 (bottom right), 17, 21

Editor: Meg Gaertner
Series Designer: Jake Slavik

Library of Congress Control Number: 2019942793
Publisher's Cataloging-in-Publication Data
Names: Gaertner, Meg, author.
Title: Measuring volume / by Meg Gaertner
Description: Minneapolis, Minnesota : Pop!, 2020 | Series: Let's measure | Includes online resources and index.
Identifiers: ISBN 9781532165573 (lib. bdg.) | ISBN 9781532166891 (ebook)
Subjects: LCSH: Volume (Cubic content)--Juvenile literature. | Volume perception--Juvenile literature. | Thickness measurement--Juvenile literature. | Measurement--Juvenile literature. | Mathematics--Juvenile literature.
Classification: DDC 530.813--dc23

Hello! My name is
Cody Koala

Pop open this book and you'll find QR codes like this one, loaded with information, so you can learn even more!

Scan this code* and others like it while you read, or visit the website below to make this book pop.

popbooksonline.com/measuring-volume

*Scanning QR codes requires a web-enabled smart device with a QR code reader app and a camera.

Table of Contents

Chapter 1
What Is Volume?. 4

Chapter 2
Units of Measurement . . . 10

Chapter 3
Tools for Measuring 14

Chapter 4
Measure It! 20

Making Connections 22
Glossary. 23
Index 24
Online Resources 24

Chapter 1

What Is Volume?

Volume is the amount of space a 3D object takes up. The object can be **liquid** or **solid**. For example, a swimming pool can hold a certain volume of water.

Watch a video here!

People use volume in their daily lives. For instance, people fill up their cars with gasoline. Most cars have a gas tank. The tank has a volume. That volume is how much gasoline it can hold.

People also use volume when cooking with some **recipes**. They measure the volume of different **ingredients**.

Chapter 2

Units of Measurement

Most **recipes** in the United States use the basic unit of the cup. One cup holds 8 **fluid ounces**.

Learn more here!

Bigger units include the pint, the quart, and the gallon. One pint is 2 cups.

One quart is 4 cups. One gallon is 16 cups.

In the United States, milk mostly comes in gallon or half-gallon containers.

Chapter 3

Tools for Measuring

When cooking, people can use **measuring cups** and measuring spoons to find volume. Each cup or spoon holds a specific volume.

Learn more here!

For example, a **recipe** may call for 1 cup of sugar. People can fill the 1-cup measuring cup with sugar.

> Measuring cups come in many sizes. These sizes include 1 cup, ½ cup, ⅓ cup, and ¼ cup.

People can also use measuring jugs. These jugs are made of glass or plastic. They have lines along their sides. The lines mark the different units.

Chapter 4

Measure It!

To measure an unknown **liquid** volume, pour the liquid into a measuring jug. Find the line that is closest to the top of the liquid. That line gives the volume.

Complete an activity here!

Making Connections

Text-to-Self

Have you ever seen someone measure volume? What tool did that person use?

Text-to-Text

Have you read other books about volume? What did you learn?

Text-to-World

A recipe asks for 3 ½ cups of water. Morris has a 1-cup measuring cup and a ½-cup measuring cup. How can Morris measure out the right amount of water?

Glossary

fluid ounce – a unit of volume used for measuring liquids.

ingredient – a substance used in a recipe.

liquid – a substance that has a constant volume but no definite shape.

measuring cup – a cooking tool that measures the volume of liquid or solid ingredients.

recipe – a list of directions for making food.

solid – a substance that has a definite volume and shape.

Index

cooking, 8, 14

cups, 10, 12–13, 14, 16, 21

fluid ounces, 10

liquid, 4, 20

measuring jugs, 18, 20

measuring spoons, 14

recipes, 8, 10, 16

solid, 4

Online Resources

popbooksonline.com

Thanks for reading this Cody Koala book!

Scan this code* and others like it in this book, or visit the website below to make this book pop!

popbooksonline.com/measuring-volume

*Scanning QR codes requires a web-enabled smart device with a QR code reader app and a camera.